MATH LYRICS FOR THE AGES 13-103

Naira R. Matevosyan, L'Auteur Libraire
Copyright © 2017, ISBN: 978-1973720331

EIGHT POEMS

Trapezoid	3
Sphere	7
E	14
Infinity	22
Chaos	27
The Pick's theorem	38
The birthday paradox	43
The Riemann's zeta	47

Area = 1/2 x h x (b1 + b2)

TRAPEZOID

Put aside that Android
As we study trapezoid !

If your roof - front has four verges
In its two-dimensional (2D) plane,
Two are tilted, two are straight,
Tilted are the legs (L) - congruent and even,
Straights are the bases (B) - parallel, uneven;

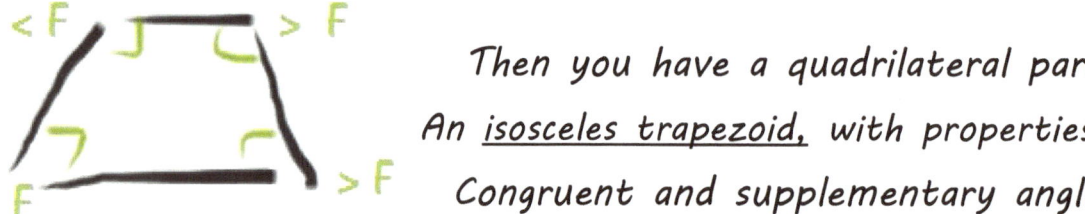

Then you have a quadrilateral parallelogram,
An <u>isosceles trapezoid,</u> with properties like these:
Congruent and supplementary angles (< F),
Congruent diagonals (D) that never meet at a right angle,
And a midsegment (M) half the sum of the bases (b1 + b2).

Can you solve this?

$M = 1/2 \; (b1 + b2)$

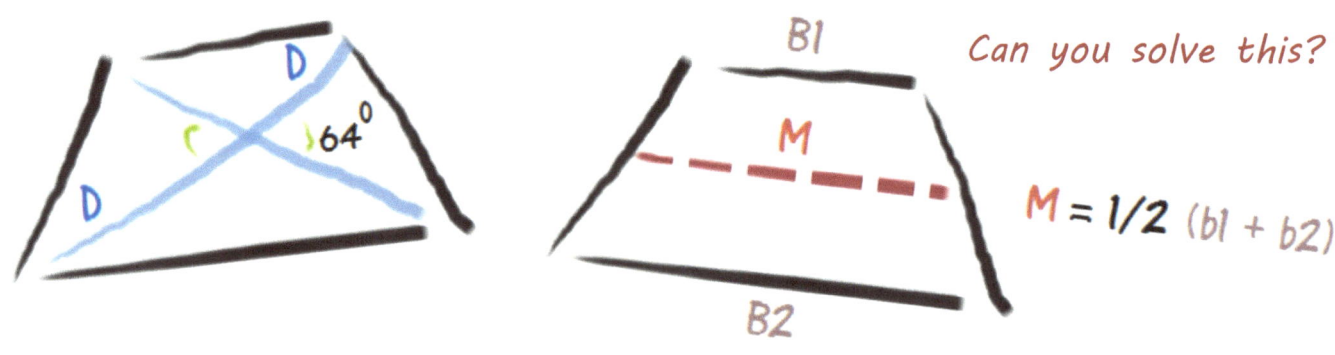

Sure you can, once you remember that
A trapezoid is isosceles if - and only if - it has
Parallel bases and congruent base angles,
Congruent diagonals and supplementary opposite angles.

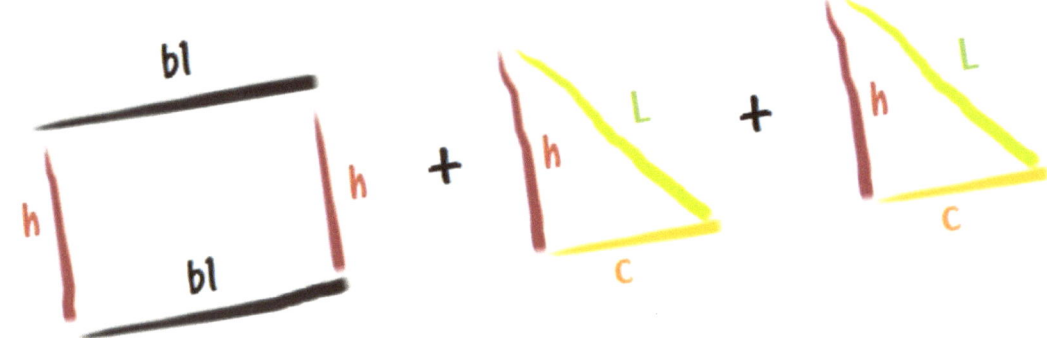

And so, you'll make a trapezoid
By adding to a quadrange (rectangle or square)
Two right triangles with equal hypotenuses (L)
And opposite sides just like the height (h) of your rectangle or square.

h - opposite side
L - hypotensus
C - adjucent side

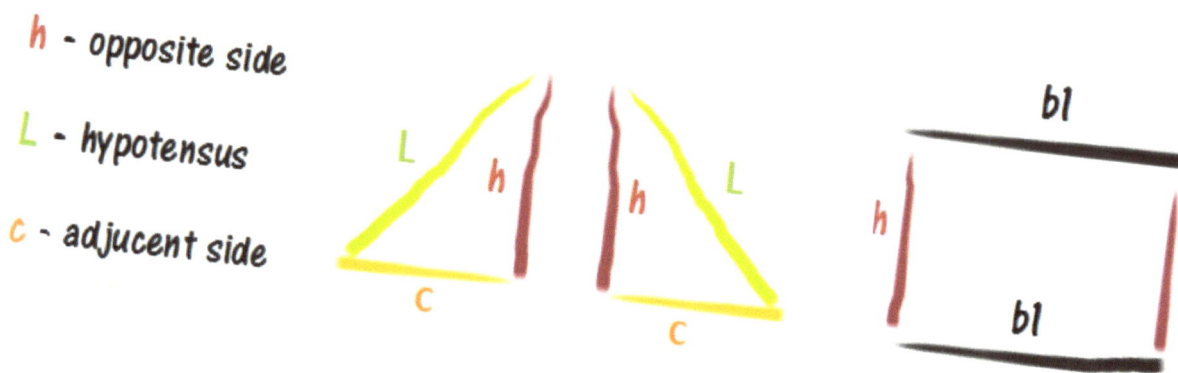

The rest will build up itself!
The opposite sides of triangles (h) will match the height of trapezoid,
The adjacent sides of triangles (c) will adjust the trapezoids's base (b2),
Where b2 = b1 + 2c. That's the case!

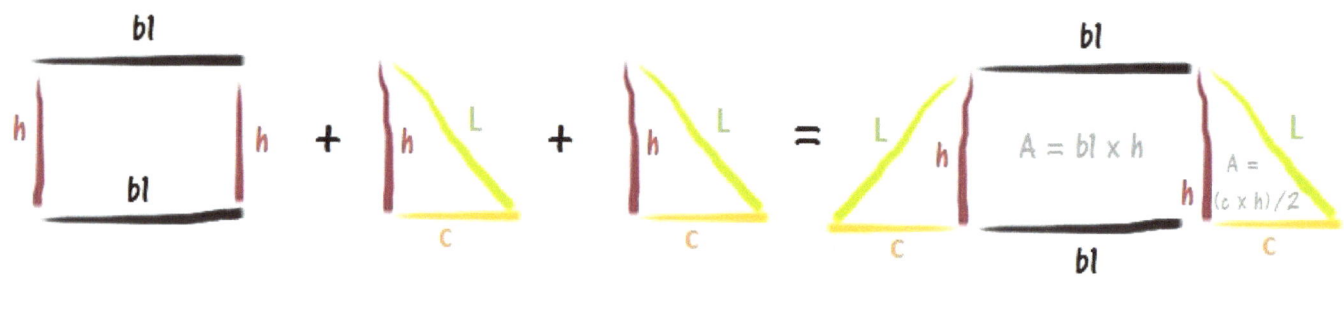

Now, grab your Android as you have an Euclidean queue.
If the area of rectangle is b1 x h
And the area of each right triangle is (c x h) / 2
Then the area of two triangles is c x h
And the area of trapezoid is b1 x h + c x h, or h x (b1 + c).

The standard equation for trapezoid is (A) = h/2 x (b1 + b2)
And if your equation (A) = h x (b1 + c) is true
Then h/2 x (b1 + b2) = h x (b1 + c), or (b1 + b2) x 1/2 = b1 + c
Or b1 /2 + b2 /2 = b1 + c, or b2 = b 1 + 2c
Proving again that A = h x (b1 + c) or A = h/2 (b1 + b2) is true.

SPHERE

Space isn't made of points, lines, or cords,
Rather points, lines, and cords are in our minds.
Scientists and dreamers know how to browse
To a space where the image-power grows.

As you wander across the galaxy
You can't help it but to wonder
Whether you're staring at a luminous orb,
Or a genuine, telestrial globe?

By dint of fear of Infinity and Oneness
You structure a space to confine
A theory that all is One;
And there, you draw the line.

You paint dots and lines on a 2D plane
To show that sphere is a set of points
In a 3D Euclidean space (R^3)
With equidistant radii from the core.

Twice a radius (r) is the diameter (d),
Formed by connecting pairs of points
On the reverse sides of the globe,
Also known as the antipode (p').

Once you know the sphere basics -
There's something you mustn't miss:
Sphere has a diverse meaning
For the geometers and topologists.

Geometers see it as a 2D circle,
A number of coordinates in a plane.
Topologists refer to the surface itself,
A N-dimensional surface (S^n)
Where the set of points $x = (X_1, X_2, ... X_n +1)$
Are spread in E^{n+1} to satisfy $X^2 + ... + X^2_{n+1} = 1$.

While geometers name the sphere-surface a "3-sphere"
Topologists denote it as S^2, referring to the "2-sphere."
Regardless the diversity in indexing and conventions
The term "sphere" means a two-dimensional surface.

Yet, the colloquial meaning of a spheric interior isn't about "sphere,"
Rather the spheric interior, the "ball," is what forms the volume.
You can find the surface area (S) and the volume (V) by solving
$S = 4\pi R^2$, and $V = 4/3 \pi R^3$, and here is why:

> The clue is in *Cavalieri principle*.
> Imagine, you have two solids (cone and cylinder) with identical heights
> And you cut them with parallel planes.
> If the cross-section areas are indentical, so too are the solid volumes.

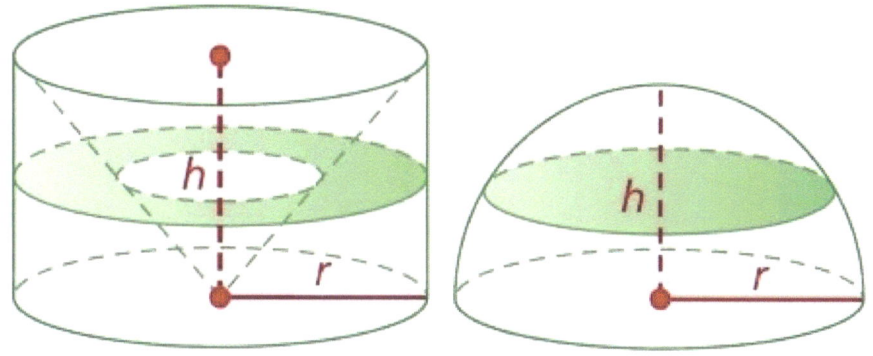

> We know, the volume of a cone is the 1/3 of a cylinder's
> And if both have the same bases and heights
> *Cavalieri* method helps you find the sphere's volume
> By comparing a hemisphere to a cylinder with an inscribed cone.

Draw a cylinder with a radius R and height R
Which contains an inverted cone with a base-radius of R -
Coinciding with the top of the cylinder with a height R.
Now, put next to it a hemisphere with a radius R.

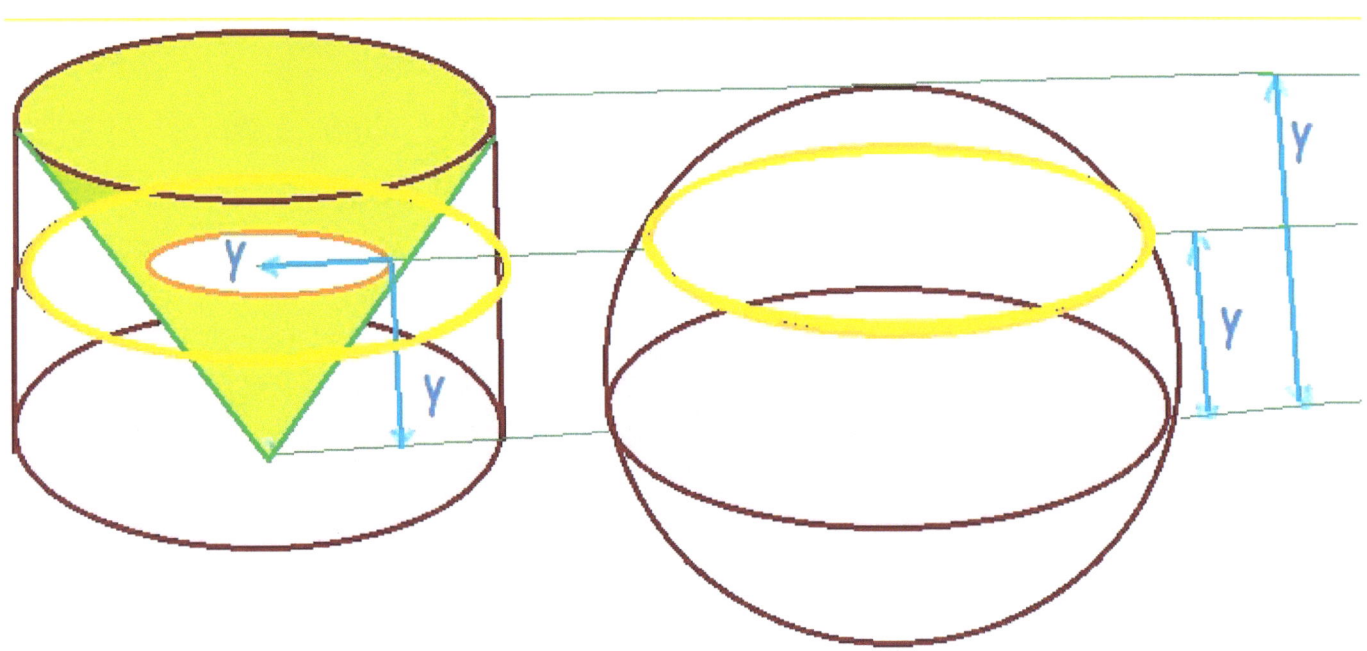

The cross section of each will have areas of $\pi(R^2 - y^2)$.
Per *Cavalieri,* if a cylinder and a cone have identical heights,
The cylinder's volume is πR^3 and the 1/3 of it is the cone's.
The remaining $2/3 \pi R^3$ is the hemisphere's, and the sphere's is $4/3 \pi R^3$.
That is what the geometers suggest.

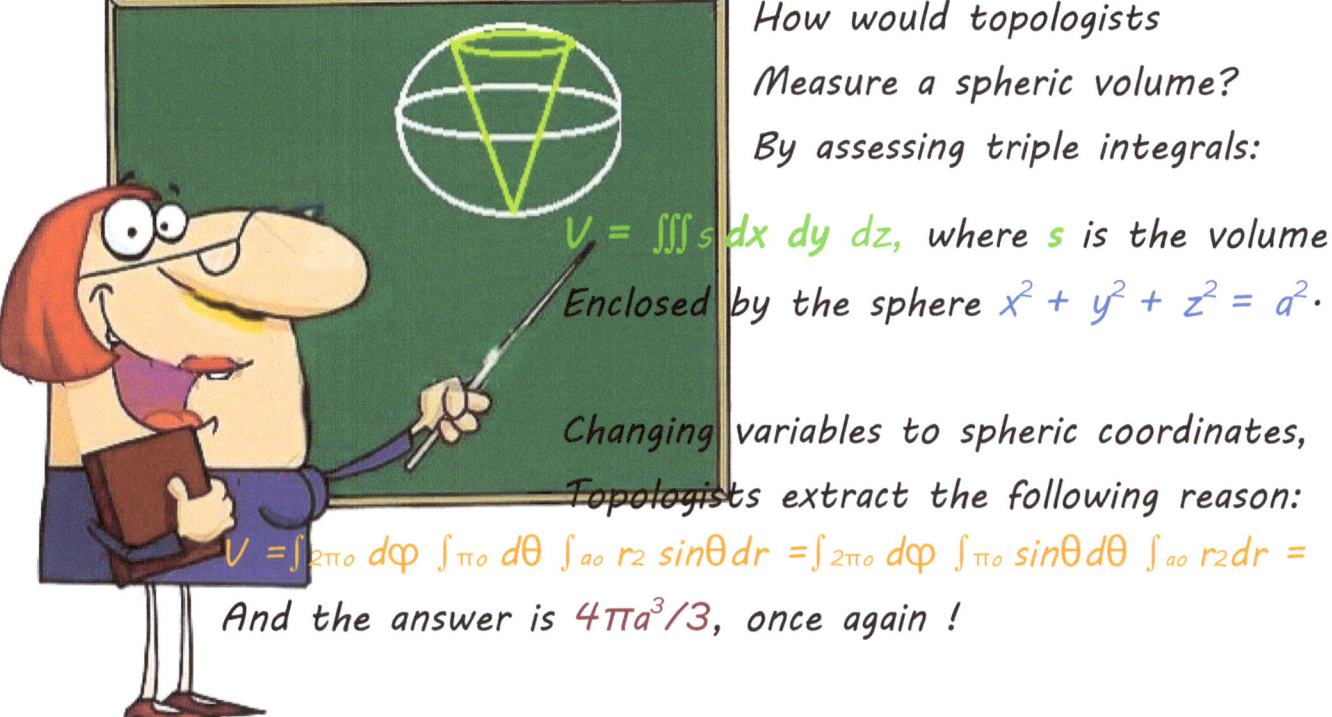

How would topologists
Measure a spheric volume?
By assessing triple integrals:
$V = \iiint_s dx\,dy\,dz$, where s is the volume
Enclosed by the sphere $x^2 + y^2 + z^2 = a^2$.

Changing variables to spheric coordinates,
Topologists extract the following reason:
$V = \int_{2\pi 0} d\varphi \int_{\pi 0} d\theta \int_{a0} r^2 \sin\theta\, dr = \int_{2\pi 0} d\varphi \int_{\pi 0} \sin\theta\, d\theta \int_{a0} r^2 dr =$
And the answer is $4\pi a^3/3$, once again !

Put simply, when you revolve $y = \sqrt{r^2-x^2}$ about the x - axis
Then you get the volume of a sphere.
To prove it, form a disk with height $f(x)$
And find the area of it.

The area of the red disk is πr^2 or $\pi f^2(x)$,
Or we could safely say $\pi \sqrt{r^2 - x^2} = \pi (r^2 - x^2)$
At any point-x, between $x = -r$ and $x = +r$.

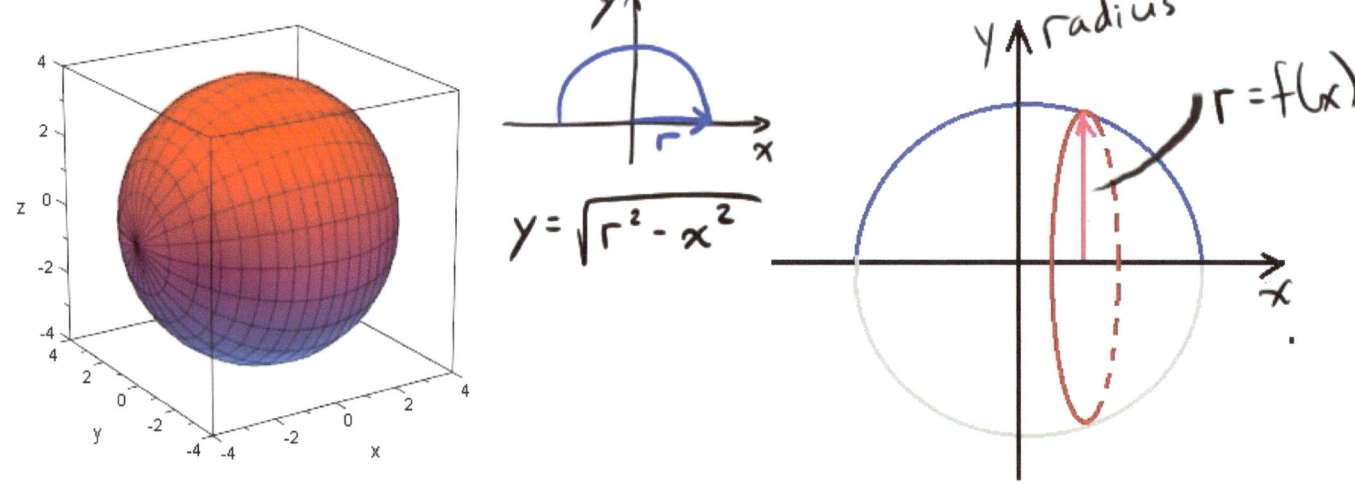

To find the volume of a sphere,
You have to add the areas of infinitesimally thin disks
As x goes from -r to +r.

    To make it work as a bliss, try this:

$\int_{-r}^{r} \pi \sqrt{r^2 - x^2}\, dx = 2\pi \int_{0}^{r} r^2 - x^2\, dx = 2\pi(r^2 x - 1/3 x^3)|_{0}^{r} =$

$= 2\pi(r^3 - 1/3 r^3) = = 4\pi r^3 / 3$

Archimedes was the first to derive an equation
In terms of expressed π and the sphere's circular cross section:
V_{sphere} / ($V_{circumscribed\ cylinder}$ - V_{sphere}) = 2.

For a sphere with radius R
Centered at the origin,
Cartesian coordinates
$x^2 + y^2 + z^2 = R^2$ show
That any cut is a circle,
With the largest section
Crossing the diameter.

The case is different
With the ellipsoid,
$x^2/a^2 + y^2/b^2 + z^2/c^2 = 1$
And with the spheroid,
$(x^2 + y^2)/a^2 + z^2/c^2 = 1$.

Also remember:
A sphere with a center at the origin can parametrically be seen as
$u = r\cos\theta$, where $x = \sqrt{r^2 - u^2}\cos\theta$, $y = \sqrt{r^2 - u^2}\sin\theta$, and $z = u$,
And where θ runs from 0 to 2π, and u runs from $-r$ to $+r$.

Did you, or did not, enjoy our sphere - talk,
Do your best to enjoy your life on our one-and-only sphere ...

<u>e</u> is a kid when compared to its rival π (3·14)·
While π is more august, with history backdating to the Babylonians
<u>e</u> was implied with logarithms in 1614 by John Napier,
And born in 1720 to its Swiss father, Leonhard Euler·

    Whether it's population, economics, space, or biology,
    "Growth" invariably involves the vibrant and youthful <u>e</u>·
    <u>e</u> is a constant, an irrational number like π
    With an unknown exact value, somewhere next to 2·718281828·

So, why *e* is that special?

e isn't a number picked out at random,

It rather is one of the greatest mathematical constants

With its chief relation to the growth. Suppose:

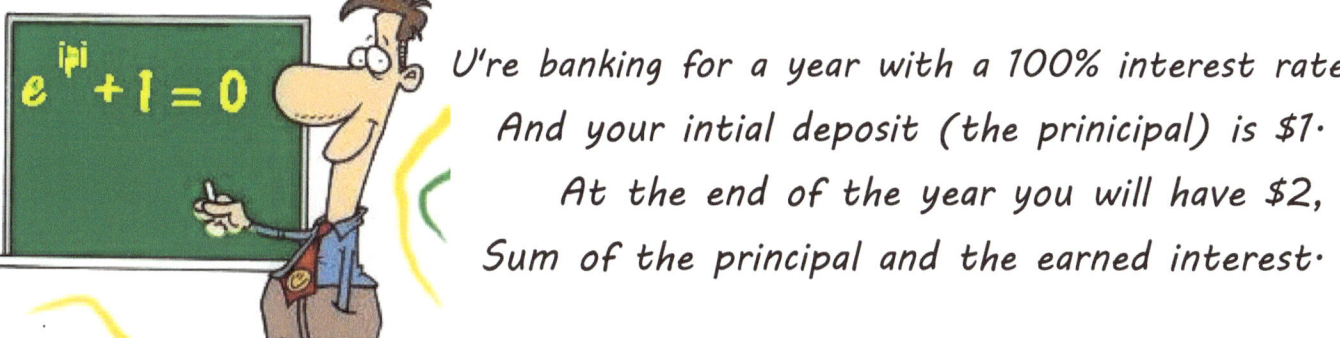

U're banking for a year with a 100% interest rate

And your intial deposit (the prinicipal) is $1.

At the end of the year you will have $2,

Sum of the principal and the earned interest.

Now, assume your interest rate is halved to 50%

And applied for each half of the year separately.

Six months after you'll earn an interest of $1.50,

And by the end of the year its half ($ 0.75) will be added.

Your annual interest rate will grow from $1 to $2.25.

In other words, with your $10,000 investment

You will get an interest of $2,250 instead of $2,000,

As by compounding for every ½ year you earn additional $ 250.

Next, suppose the year is split into four quarters
With 25% of interest applied for each;
Carrying out the same calculation
You'll find that your interest has grown to $2.44141.

So, your money is growing like the plant in p.14
And your investment will be more advantageous
If you split up the year to shorter bouts,
Applying smaller interest rates to those.

Certainly, the most realistic compounding period is the day
And that is what the banks grasp to thrive;
The morale of this story is in "_e_" -
That makes your $1 flourish through compounding rise.

Is that a good thing or a bad thing?
It all depends: if you've invested and you're saving
_e_ will make you a billionaire;
But if it's a loan, _e_ will compound it and throw you to the couloir.

e can be both fortune and terror,
A recipe for triumph or fiasco.
But you can wisely manage *e*
Once you know its anatomy.

If the value of *e* is 2·718281828459045233536···
The best fitting fraction will be 87/32.
Accordingly, the serios of *e* explansion would be
e = 1 + 1/1 + 1/(2 x 1) + 1/(3 x 2 x 1) + 1/(4 x 3 x 2 x 1) ···

$2.6 < e < 2.9$	$n = 10$
$2.70 < e < 2.73$	$n = 100$
$2.717 < e < 2.720$	$n = 1000$
$2.7181 < e < 2.7184$	$n = 10000$
$2.71827 < e < 2.71830$	$n = 100000$
$2.718280 < e < 2.718283$	$n = 1000000$
$2.7182817 < e < 2.7182820$	$n = 10000000$
$2.71828181 < e < 2.71828184$	$n = 100000000$
$2.718281827 < e < 2.718281830$	$n = 1000000000$

It's handy to use the factoral notation with exclamation mark:
$e = 1 + 1/1! + 1/2! + 1/3! + 1/4! + 1/5! \ldots$
It shows, that e has a distinct pattern,
Which makes it more "settled" than π.

Want to remember e's 15-place approximation value by heart?
You have no problem. Follow this mnemonics!
2, and Andrew Jackson (the 7th president) was elected in 1828.
His "Old Hickory" picture has frames of 18 and 28 inches.

You already have this much expansion: 2.718281828.

What about the remaining 45904523536 ?
Remember, diagonals split squares into triangles
Making these angles: 45-90-45.

Still have 235336 left?
Don't you worry. Citing Holy Bible, Luke 23:53:
"And he took it down, wrapped it in a linen,
And laid Him in a tomb cut into the rock…"

Scaled. Initial measures 18 x 28 in

Still have 36 to go! Here we go:
Jewish faith holds that the number 36 is special,
As light created by God on the First Day shone for exactly 36 hours
Before it was replaced by Sunlight created on the Fourth Day.

Hint: Time measures where strikingly longer in Biblical / Hassunah period.
For example, Abraham lived 175 years and fathered Isaac at his 99,
Which makes it clear why the 12-hour sunshine was thrice longer then.
Nontheless, Bible can still help us build a math mnemonic!

Next line: If Holy Bible is "transcendental," what about e ?
Can an irrational number be transcendental?
By strict algebraic standards, e is a conjecture, as the proves suggest
It's impossible for both e and its power (e^2) to be transendental.

> The connections between e and π are truly fascinating!
>
> While the values of e^{π} and π^e are close,
>
> It takes a fast cheating to show that $e^{\pi} > \pi^e$
>
> As your Android shows that e^{π} = 23.14069, and π^e = 22.45916.

The number e^π is known as Gelfond's constant,
Named after Alexander Gelfond, a Russian mathematician.
That number e^π is transcendental.

Lesser is known about π^e. It hasn't been proved yet to be irrational.

e is an earth-suttering identity,
As it is present
In the most remarkable
Mathematical reasons.

When we think of the famous math figures
We think of 0, 1, π, *e* and the imaginary number $i = \sqrt{-1}$.

How could be that $e^{i\pi} + 1 = 0$?

It truly is! A result attributed to Leonhard Euler.

Withal, e is unavodable!
If a talented author would think of writing an e-less novel,
He should probably hide himself under a pseudo pen-name.
Yet, it's hard to believe that there is a mathematician
Setting out to write an e-less textbook, or even is able to do so.

e has inspired generation of mathematicians: from Joseph Liouvlle,
Who showed that e is not the solution of any quadratic equation
To Charles Hermite, who proved that e is transendental,
Before Ferdinand von Lindenmann would adapt the idea to show
That π too was transcendental with a much higher profile.
So, you're not the only one "going mental" !

$$e \approx \frac{1}{1!} + \frac{1}{2!} + \frac{1}{3!} + \frac{1}{4!} + \ldots$$

$$e \approx \sum_{n=1}^{\infty} \left(1 + \frac{1}{n}\right)^n$$

INFINITY

How big is infinity? Can we transfine it?
Assume, for every giant number such as 10^{1000}
There is always a greater one such as $10^{1000} + 1$.
This is a traditional view of infinity, with numbers marching on forever.

Yet, Georg Cantor, a German mathematician, had a different concept
Which had to do with a primitive notion of counting.
Cantor's theory involved number sets, like N = {1, 2, 3, 4, 5, 6, 7…},
And subsets, like O = {1, 3, 5, 7…} and E = {2, 4, 6, 8…}.

The inquiry was whether there are equal counts of odds and evens?
We'd say "yes" and Cantor would agree with our concept of mating
Where "half the whole numbers are odd and half are even."
There rises another que: "are there same numbers of odds and evens?"

Not necessarily, when it comes to the odds and primes.
Yet, there is one-to-one correspondence between the sets of N and E.

N:	1	2	3	4	5	6	7	8	9	10	11
E:	2	4	6	8	10	12	14	16	18	20	22

This allows us to say, "there's the same count of whole and even numbers."
It files right in the face of the "common notion"
A concept developed by the ancient Greeks in "Euclid of Alexandria's" -
A textbook of "elements" where "the whole is greater than the part."

The number of elements in a set is called "cardinality."
It's the measure of the size of a set. For our set of N,
Cantor used the symbol "aleph"(ℵ) or "aleph nought"
Which would note our set as card(N) = card(O) = card(E) = card ℵ₀

So, any set available for one-to-one correspondence with N
Is "countaby infinite." That means, whe can sort the set.
Are the fractions contably infinite?
Let's check -- by moving to a two-dimensional plate.

Build a set of whole numbers (N)
And mate it to the subset of fractions (Q).

Can we devise a list including both positive and negative fractions
Mated one-to-one with the whole numbers? No, that's impractical.

Yet, you can do something about it
If you write all fractions with 2 as denominator
Followed by a row of fractions with 3 as denominators
And where repeating numbers and integars (4/2, 6/3) are ommited:

1	-1	2	-2	3	-3	4	...
1/2	-1/2	3/2	-3/2	5/2	-5/2	7/2	...
1/3	-1/3	2/3	-2/3	4/3	-4/3	5/3	...
1/4	-1/4	3/4	-3/4	5/4	-5/4	7/4	...
1/5	-1/5	2/5	-2/5	3/5	-3/5	4/5	...
...	...	...	...	...	...	...	...

Holding in this fashion, you will devise a one-dimensional list
And by choosing a devious zig-zagging route, you will locate the mates.
Then you'll conclude that *"the set of fractions Q is countably infinite"*
Signing it as *card(Q)* = $\aleph_0$

1 —	2	6 —	7	15 —	16	25 —
3	5	8	14	17	24	etc
4	9	13	18	23	etc	etc
10	12	19	22	etc	etc	etc
11	20 —	21	etc	etc	etc	etc

What about the irrational numbers, like $\sqrt{2}$, e or π
That fill the "gap" to give the continuum of the real number line R?
Cantor proved that any attempt of "filling the gap" between 0 and 1
With the help of real numbers is doomed to failure.

Let's argue. You know that any number between 0 and 1 is a decimal
With both rational (like 0.5252) and irrational (like 0.5234) numbers included.
Now show your list to Cantor: $r_1, r_2, r_3, r_4, r_5 \ldots$
Cantor would probably draw a zigzag pattern bolding some numbers like:

$r_1: 0.\mathbf{a_1}a_2a_3a_4a_5\ldots$

$r_2: 0.b_1\mathbf{b_2}b_3b_4b_5\ldots$

$r_3: 0.c_1c_2\mathbf{c_3}c_4c_5\ldots$

Then he will ask: "where is the number $m = m_1m_2m_3m_4m_5\ldots$?"

Cantor will rephrase: "m_1 must differ from a_1, m_2 from b_2, m_3 from c_3"-
Working his way down the diagonal.
His m will vary from every number in your list in one-decimal place,
And so it can't be there. Cantor will be right !

In fact, no list is possible for the set of real numbers (R),
And so it is a "greater" infinite set, with higher "order of infinity,"
Than that in the Q set. The "big" just got bigger,
As claimed by Zeno of Elea or Surya Prajnapti.

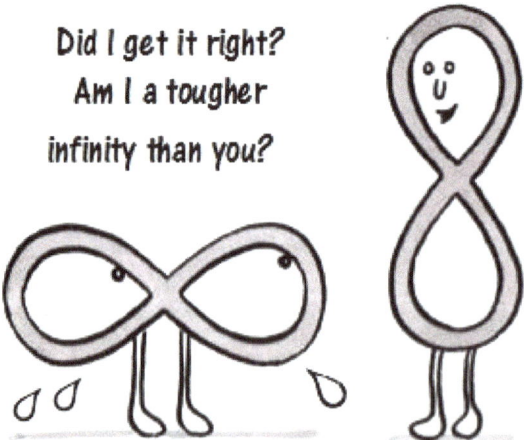

CHAOS

Is there a "theory of chaos"?
Or perhaps, chaos is the absence of a theory?
The story goes back to 1812,
When Napoleon Bonaparte was advancing on Moscow.

Tolstoy, the great Russian novelist, repeatedly reminds us
That everything is circumstantial and foresight is luck, at best.
By doing so, he relates to the Battle of Borodino and its eerie ending.
He writes, "In offering and accepting battle at Borodino,
Kutuzov and Napoleon acted involuntarily and senselessly.
And only later did historians furnish the already accomplished facts
As to the foresight and genius of the commanders
Who were the most enslaved and involuntary agents (754)".

In Moscow, at the heart of Russia, French were risking their solid army.
Aware of the liabilities, Napoleon and Kutuzov still agreed to fight.
Yet, their intricate strategies were rendered obsolete
When the actual war blazoned its start.

28

And no matter how well you're prepared for certain events,
There is only a minute-probability that the outcome will be as expected.
Historians aver that "the Russians purposely drew the French to Borodino,"
But Tolstoy claims that "they ended up there by a chance."

Some historians argue that "the French lost because Napoleon had a cold."
Tolstoy laughs, if this were true then the Russian victory
"... would be due to the valet, who didn't give Napoleon his rain boots.
To the question of what constitutes the cause of historical events,
a different answer presents itself, which is that the course of world events
is predestined from on high, depending on the coincidence of all wills of the
game players; and that Napoleon's influence on the course of such events
is only external and fictitious (784)."

The French Emperor's compatriot, Marquis Pierre-Simon de Laplace,
Published an essay on determanitsitc universe, claiming
"If at one singular instant, positions, velocities,
And forces applied on all obsjects of universe were known,
Then these quantities would be easily determined and predicted."

Yet, the real world
Is more intricate than that.
Small discrepancies in initial conditions
Mean discrepancies in the outcomes.

The <u>Butterfly Effect</u> holds:
*"If a fine weather
Is predicted in Utah on day-U,
And if butterfly flaps its wings
In Arizona,
This could actually
Presage storms on Utah,
Because that flapping of the wings
Slightly changes the air pressure causing a weather pattern
Completely different from the one forecasted."*
Hey, remember the Facebook's impact on the US's 2016 election?

Emperor Bonaparte before the Battle of Baradino

With all repects, from "metheorology" to math,
The discovery of butterfly effect in itself happened by chance, in 1961.
The meteorologist Edward Lorenz at MIT took a coffee break
And when he was back to his desk, something abrupt had happened.

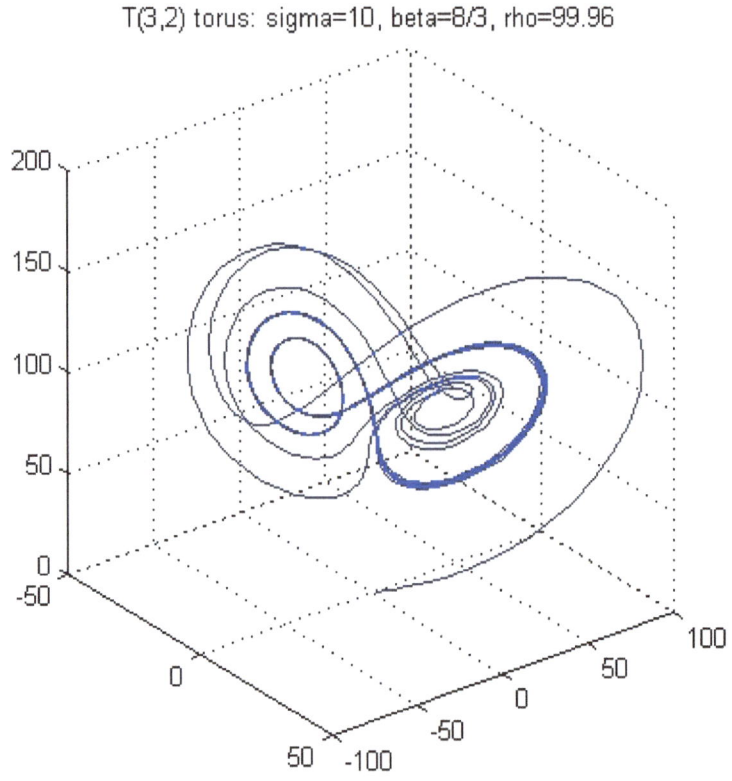

The wheather plot he had left
For the coffee break
Had undergone unthinkable turns
Under the same initial values
That Lorenz had used
To predict otherwise.

Would he dump his computer,
For changing his 6 decimal places
Into 3 decimal values?
Lorenz didn't panic. Instead,

He developed a concept of
The "butterfly effect,"
Migrating his intellectual interests from meteorology to mathematics!

Below we test the idea with a simple mechanical experiment.
Drop a ball-bearing through the opening in the top of a pinboard box
And see, it will progress downwards – being deflected one way or the other
By different pins it encounters on route
Until it reaches a finishing slot at the bottom.

You might then attempt to let another identical ball-bearing
Go from the very same position, with exactly same velocity.
If you could precisely do this, then Marquis de Laplace would be right,
As the path followed by the ball would exactly be the same.

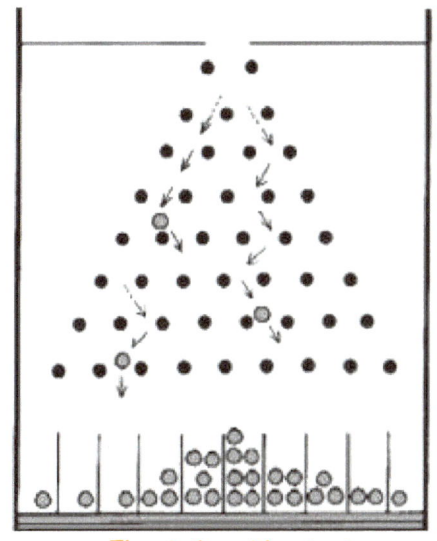
The pintboard box test

If the first ball dropped into the 3rd slot,
Then so would the second ball.
Yet, in the reality, you can't let the ball go
From exactly same position
With exactly same velocity and force.
There will be a slight difference
Which you'll not even be able to measure.

In the result, the ball-bearing may take a very different route,
Ending up in a different slot.

A simple pendulum is one of the best mechanical systems to reflect on.
The pendulum gradually loses energy as it swings back and forth.
The displacement from the bob's vertical and the angular velocity
Decreases until it is eventually stationary.

The bob's movement can be plotted
In a phase diagram, where the horizontal axis
Stands for the angular displacement,
And the vertical one for the velocity.

The point of release is plotted at the point A
On the positive horizontal axis.
At A the displacement is at a maximum
And the velocity is zero.

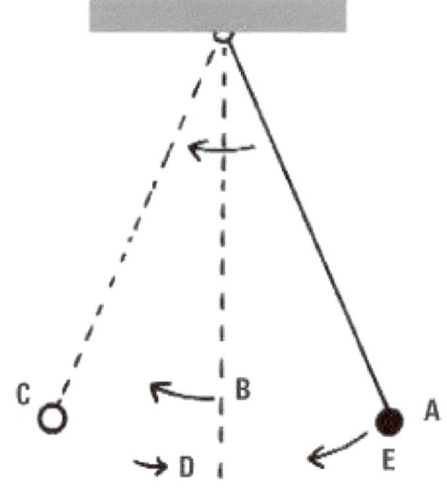

The free pendulum

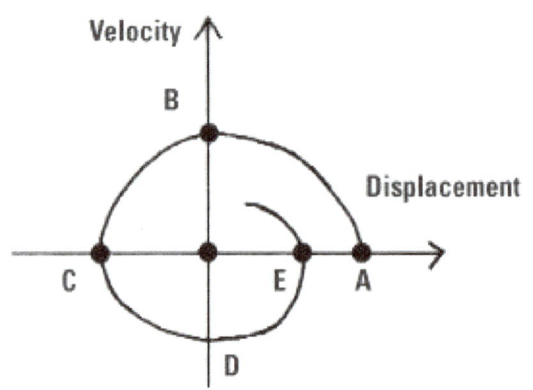

Phase diagram for a simple pendulum

As the bob moves through the vertical axis
The displacement is zero
And the velocity is at a maximum,
As plotted on the phase diagram, point B.

At C, at the other extremity
To where the bob swings,
The displacement is negative
And the velocity is zero.

The bob then swings back through D
While it moves in the opposite direction,
During which time its velocity is negative –
When it completes one swing at E.

In the phase diagram this all is represented by a rotation through 360°
But because the swing is reduced the point E, it is shown inside A.
As the pendulum swings less and less, it spirals into the origin
Until it comes to the final rest.

That's not the case with the double pendulum
Where the bob is at the end of a jointed pair of rods.
If the displacement is small
The motion of the double pendulum is similar to that of simple sundial.

Where the displacement is large, the bob swings, rotates, and lurches
And the displacement about the intermediate joint is seemingly random.
If the motion is unforced, the bob will come to rest
But its motion curve will be far from the spiral of a single pendulum.

Now, where's the chaos? Chaos is when a deterministic system
Appears to generate a random behavior.
Let's tackle another case, as repeating and iterative as $a \times p \times (1 - p)$
Where p stands for the population, measured as a proportion on scale 0-1.

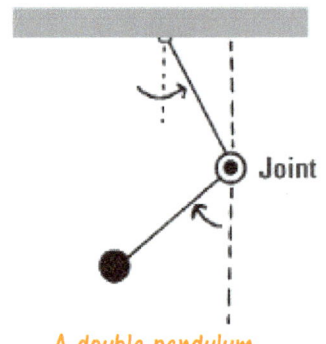
A double pendulum

The a - value must be somewhere between 0 and 4
To ensure that the p-value stays in the range 0 - 1.
Let's model a population count as $a = 2$
With a starting value of $p = 0.3$ at time = 0.

To find the population at time = 1,
We feed $p = 0.3$ into $a \times p \times (1 - p)$ to produce 0.42.
Just by using a basic calculator we can iterate this operation,
This time with $p = 0.42$, to give us the next figure (0.4872).

Progressing in this way, we find the population at later times.
In this case, the population quickly settles down to $p = 0.5$
At the a - value of less than 3. If now we choose $a = 3.9$,
The maximum permissible a, p 0.3 will not settle, it'll oscillate wildly.

This is because the *a* - value is in the 'chaotic region' as *a* > 3.57.
Now, if we choose an initial *p* 0.29, a value close to 0.3,
For the first few steps the *p* growth will shadow the previous pattern
But then will start to diverge from it completely, as in Lorenz's case.

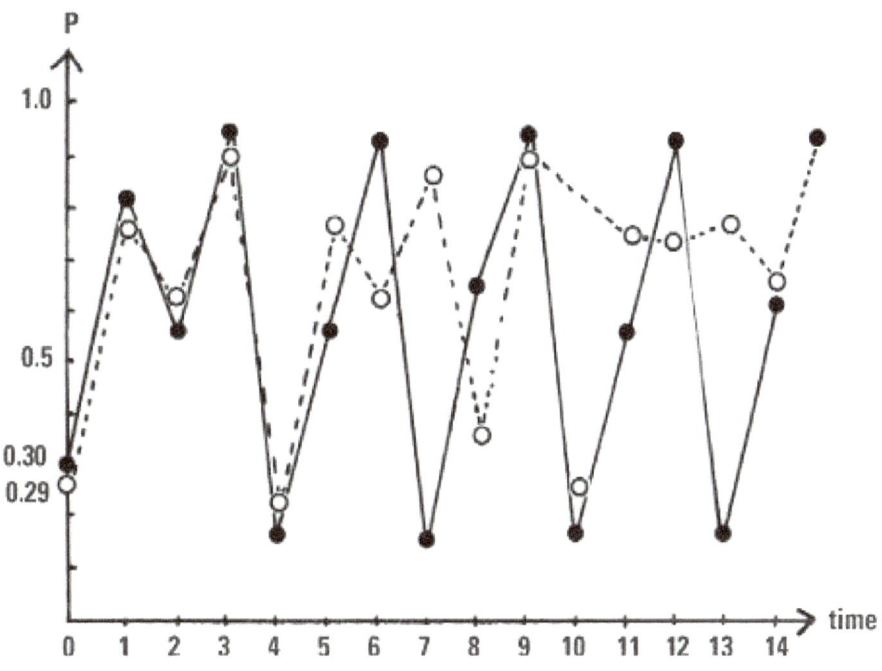

Population changing over time for a = 3.9

The theory behind the math of the weather forecast
Was tested independently, by the French engineer Claude Navier (1821)
And by the British mathematical physicist George Gabriel Stokes (1845).
Yet, the Navier-Stokes equations still challenge many math debates.

Applied to the problem of the fluid flow,
Much is known about the steady movements of the upper atmosphere.
But air flow near the surface of the Earth creates turbulence and chaos.
And there, the non-linear Navier-Stokes equations arrive to soothe.

Not conservation equations, they rather are dissipative systems
Which factor the Froude limit (dimensionless number of intertia ratio),
A speed-length ratio that reads $Fr = u_0 : \sqrt{g_0 l_0}$
Where u_0 is the flow velocity, g_0 - the external field, l_0 - the length.

In this sense, Navier-Stokes equations can't be put
Into the quasilinear homogeneous form as $y_t + A(y)y_x = 0$
These rather are existence and smoothness problems, for what
The Clay Math Institute has offered a $1million prize to solve.

Dynamic systems can be seen as strange attractions.
In the case of a simple pendulum, the attractor was the single point
At the origin that the motion was directed towards.
With the double pendulum, it is more complicated.

Yet, even in the case of the double pendulum,
There are sets of points with some regularity
To be attracted to in the phase diagram.
For randomly moving systems, such sets form fractals
Or *"strange attractors,"* with a definite mathematical structure.
But wait: are we talking about an *"organized chaos"*?

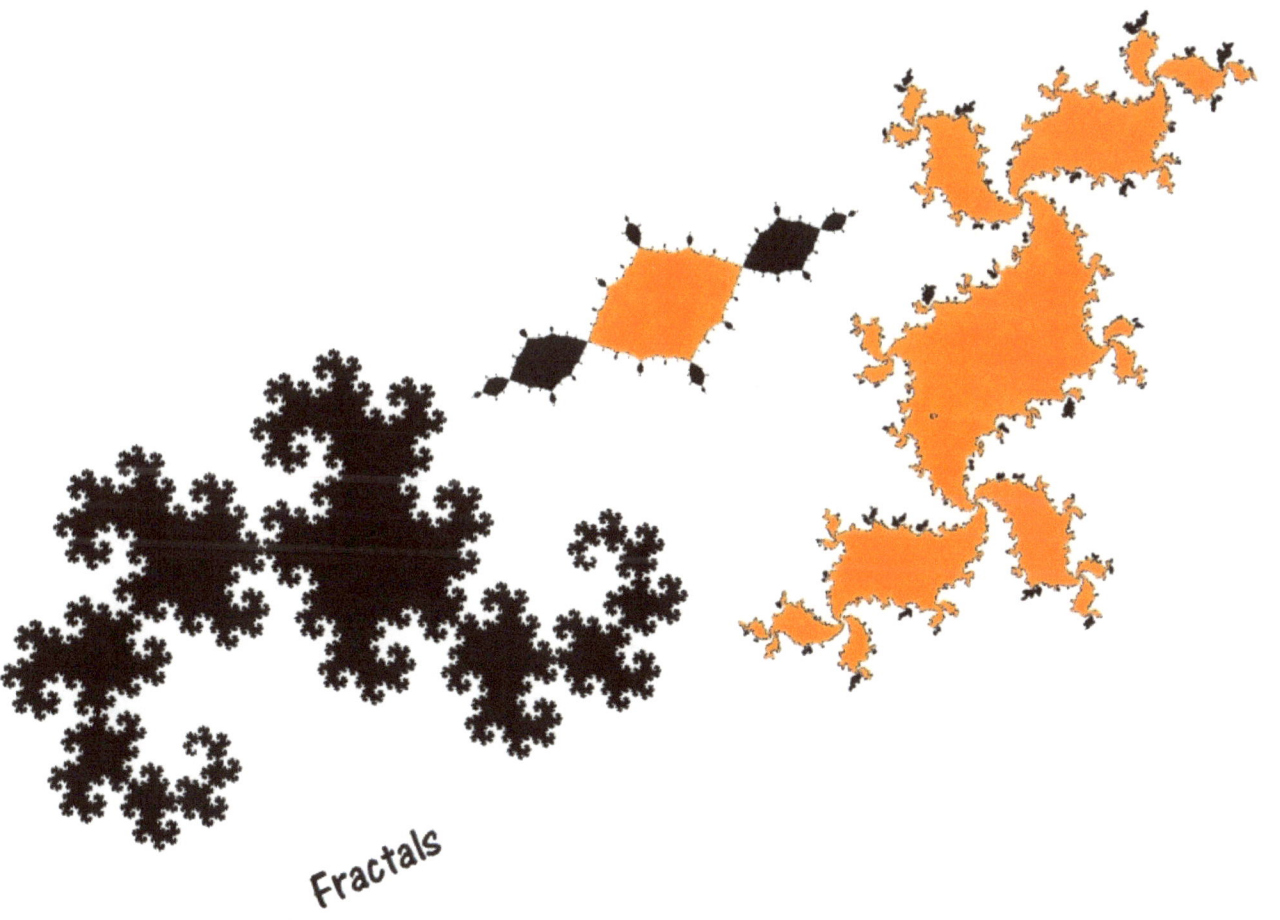

Fractals

THE PICK'S THEOREM

Georg Pick, an Austrian mathematician
Had two claims to the fame.
The first, that his reference to the
German University in Prague
Was no one else, but Albert Einstein.
The second was his authentic paper,
The "reticular geometry" (1899).

Pick's theorem helps measure
Complex areas of polygonal shapes
With a simple trick of joining points,
The coordinates of which are whole numbers.

Also known as the pinball mathematics,
It counts the number of points (b) on the boundary of the polygon
And the number of points (s) inside the area of the polygon.
In our case (p.39), these numbers are b = 9, s = 5. That's all we need!

Area = b/2 + S -1

In this case, the area is

9/2 + 5 - 1 = 8.5

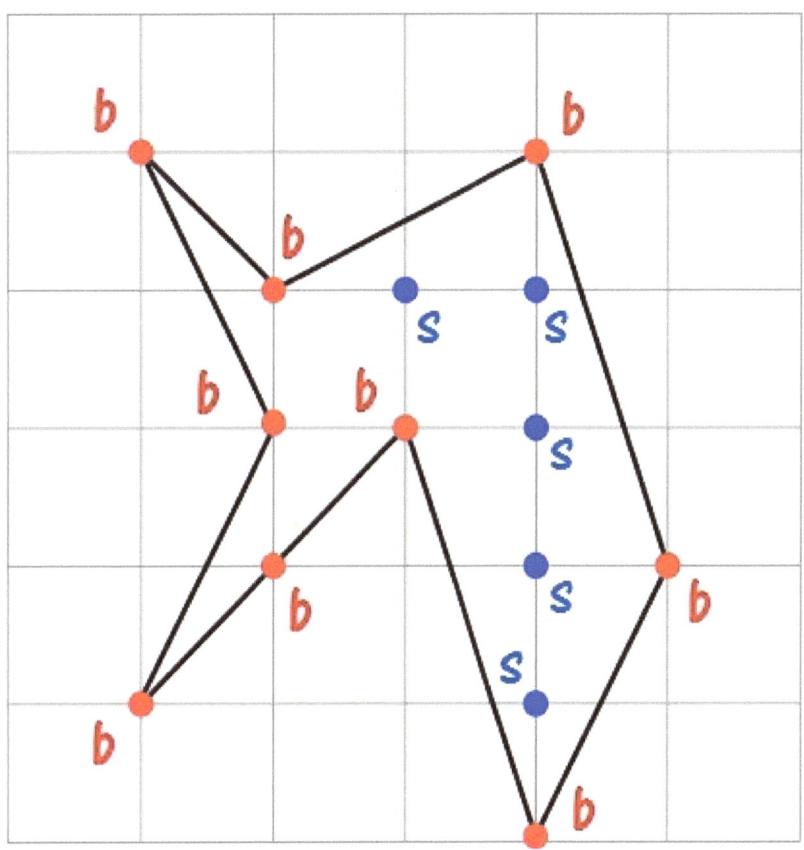

Note, the Pick's theorem has limitations.
It only applies to shapes that join discrete points of whole number coordinates and where the boundary does not cross itself.

It's only valid for polygons
That consist of a single piece and do not contain holes.
For a polygon that has holes (h)
With a boundary in the form of h + 1 simple closed curves,
A more complicated formula
s + b2 + h - 1 gives the area.

The Reeve tetrahedron shows that there's no analogue of Pick's theorem
In three dimensions that express the volume of a polytope
By counting its interior and boundary points. However,
There is a generalization in higher dimensions via Ehrhart polyhedra.

Now, let's prove it!
Consider a polygon P and a triangle T,
With one edge in common with P.
Assume, Pick's test is true for both P & T.

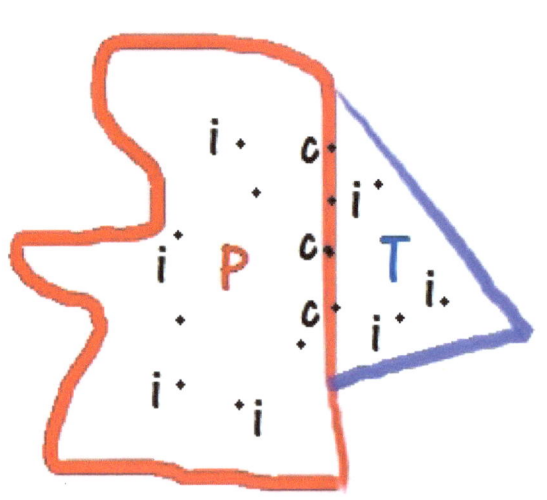

All we need is to show
That the theorem also works for PT,
A polygon obtained by adding T to P.

Since P and T share an edge, the boundary points are common (c)
Merged to the interior points (i), except the two endpoints
Which are merged to boundary points (b).
So, we have:

$$i_{PT} = i_P + i_T + (c - 2)$$
$$b_{PT} + b_P + b_T - 2(c - 2) - 2$$

And then this follows:
$$i_P + i_T = i_{PT} - (c - 2)$$
$$b_P + b_T = b_{PT} + 2(c - 2) + 2$$

Since we agreed that the theorem applies to P and T seperately,

$$Area_{PT} = Area_P + Area_T =$$
$$(i_P + b_P/2 - 1) + (i_T + b_T/2 - 1) =$$
$$i_P + i_T + (b_P + b_T)/2 - 2 =$$
$$i_{PT} - (c - 2) + [b_{PT} + 2(c - 2) + 2]/2 - 2 =$$
$$i_{PT} + b_{PT}/2 - 1.$$

So, if the theorem is true for polygons constructed from n triangles,
It is liekwise true for polygons constructed from $n + 1$ triangles.
As for the general polytopes, they can always be triangulated.
To finish the induction, it remains to prove that Pick applies to triangles.

Such verification can be done through simple short steps:
- observe that the formula holds for any unit square with vertices having integer coordinates;

- deduce that the formula is correct for any rectangle with sides parallel to the axes;
- now deduce it for right-angled triangles obtained by cutting such rectangles along a diagonal;
- now on any triangle can be turned into a rectangle by attaching such right triangles;
- since the formula is correct for the right triangles and rectangles, it also follows for the original triangle.

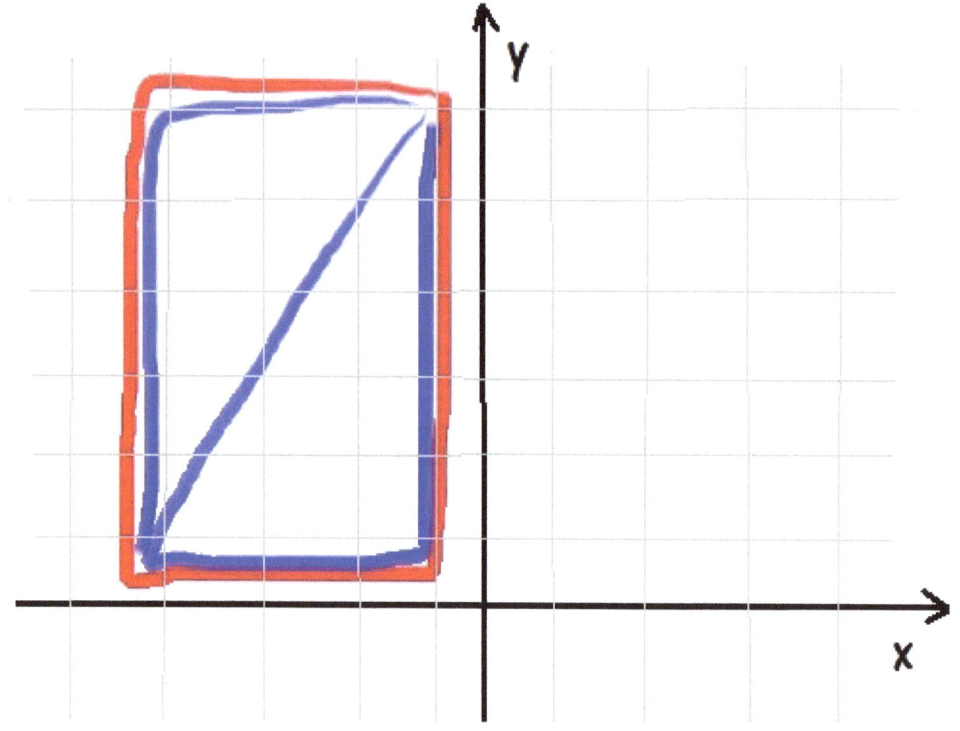

Lastly, if the theorem is true for the polygon PT and triangle T, Then it's also true for P. Prove it through the same reckoning as above.

THE BIRTHDAY PARADOX

Imagine, you're aboard the top deck of East River Ferry in New York City,
With nothing to do but to count your fellow passengers
Who commute to work in the early morning.

Reasonably, if the passengers are independent from each other,
You may safely assume that their birthdays are randomly scattered.
Let's say, there are 23 passengers per ride, including yourself.
Not much, but enough to claim that it's likely that 2 share their birthdays.

> *Would you believe it?*
> *Millions do not, but it's absolutely true.*
> *Even William Feller, a seasoned expert in probability,*
> *Found it astounding.*

Now, leave the ferry as it is too small to contain our argument.
Let's move to a larger room. How many people may gather in a hall,
To have two of them share the same birthday?
If the standard year has 365 days (n), we need at least 366 people ($n+1$).

> Known as the pigeonhole principle it suggests,
> If there are $n+1$ pigeons occupying n holes
> One hole must contain more than one pigeon.
> So, if you randomly invite 365 people to your ball,
> The chance of a common birthday is minuscule.

The more you reduce the number of people
The more unlikely is that two of them share a birthday.
Yet, the birthday problem is not a paradox.
Let's prove it, by selecting a person at random.

The probability that another person has the same birthday as your pick
Is 1/365; the probability that these two do not share a birthday
Is minus one, or 364/365. The probability that the third person
Shares birthday with these two is 2/365, and if not, then 363/365.

    And so it goes⋯
    Now, the probablity that none of these three share a birthday
    Is 364/365 x 363/365 = 132132/133225 = 0.9918

Following this flow of thought for 4, 5, 6,...n people
Unravels the birthday paradox. Back to our ferry:
If you counted 23 passengers, then the probability
That none of them shares a birthday is 0.4927, and
The probability that two of them do is 1 − 0.4927 = 0.5073.

Now, randomly pick a passenger, and give him/her a name.
Let's name him Dr Osborn. If his birthday is on April 22nd
A different question must be asked.
"How many birthdays concide with Dr Osborn's"?

The answer requires a different test.
The chance that Dr Osborn doesn't share his big-day with another person
Is 364/365, and that he doesn't share his birthday with
ANY of the $n - 1$ passengers aboard the ferry is $(364/365)^{n-1}$

Therefore, the chance that Dr Osborn does share his B-day with somebody
Will be one minus that value.
In our case of 23 passengers, it will be 0.061151, suggesting that
There is a 6% chance that someone shares his birthday with Dr Osborn's.

THE RIEMANN'S ZETA

One of the stiffest challenges in mathematics,
The *Riemann's hypothesis* remains unconquered
When *Poincaré conjecture* and *Fermat's last claim* are already deciphered.
The story starts with the addition of fractions of this kind:

$$1 + 1/2 + 1/3$$

While the answer at this stage is *1 5/6* (or *1.83*)
What will happen if we keep adding smaller and smaller fractions?

$$1 + 1/2 + 1/3 + 1/4 + 1/5 + 1/6 + 1/7 + 1/8 + 1/9 + 1/10$$

This is a *harmonic series*, a term suggested by Pythagoreans
Who claimed that a musical instrument divided by a half, a third, or a quarter
Gave more essential notes for the musical harmony.
But··· what happens to the total?

Does it grow beyond all numbers?
Or there is a barrier somewhere, blocking the infinite rise?
The trick to the answer is in grouping the terms,
And doubling the runs as we go.

Adding the first 8 terms, for example, in condition that $8 = 2^3$

$$S_{2^3} = 1 + 1/2 + (1/3 + 1/4) + (1/5 + 1/6 + 1/7 + 1/8)$$

Where S stands for the sum.

As 1/3 is bigger than 1/4 and 1/5 is bigger than 1/6,
S is greater than
$$1 + 1/2 + (1/4 + 1/4) + (1/8 + 1/8 + 1/8 + 1/8) =$$
$$= 1 + 1/2 + 1/2 + 1/2$$

Now, we can say that

$S_{2^3} > 1 + 3/2$, and more abstractly $S_{2^3} > 1 + K/2$

If K is 20, for instance, then $n = 2^{20} = 1,048,576.00$
Where the sum of series will only have exceeded by 11.

While the series is said to diverge to infinity,
This does not happen with the series of squared terms:
$$1 + 1/2^2 + 1/3^2 + 1/4^2 + 1/5^2 + 1/6^2 + \cdots$$
Here, the exponent power is 2.

The intricate phenomenon is like this:
If the power increases by a minuscule amount
To a number just above 1, the series *converges*.
And if the power decreases by a minuscule amount
To a value just below 1, the series *diverges*.
The *harmonic series* sits on the boundary between convergence and divergence.

The celebrated *Reimann zeta function* $\zeta(x)$ was assesed by Euler
Before Bernhard Riemann would warm it up observing its importance.
$$\zeta(x) = 1 + 2\square_x + 3\square_x + 4\square_x + \ldots \text{ or } \zeta(x) = 1 + 1/2^x + 1/3^x + 1/4^x + \cdots$$
But it actually was Leonhard Euler, who found that $\zeta(2) = \pi^2/6$

So it became clear, that all values of $\zeta(x)$ involve π
When x is an <u>even</u> number.
Where x is an <u>odd</u> value, the theory is far more complex.

The Riemann zeta function has an infinity of zeros,
Which means, infinity values of x for which $\zeta(x) = 0$.

In his 1859's presentation at the Berlin Academy of Sciences,
Reimann showed that all important zeros where complex numbers
Laying in the critical strip bordered by $x = 0$ and $x = 1$.

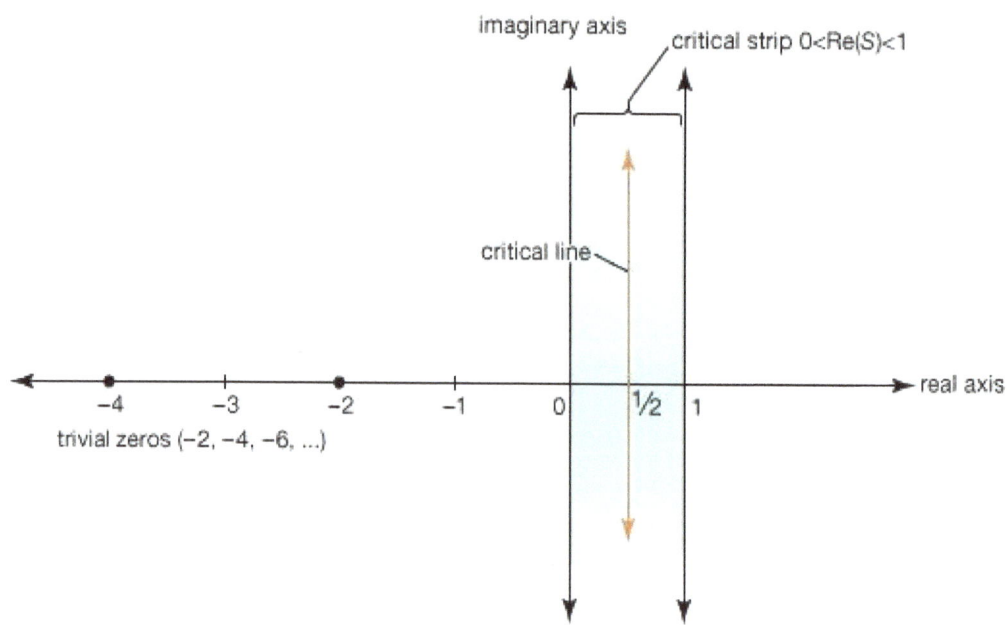

As Reimann noted then, "all zeros of the Reimann zeta function $\zeta(x)$
Lie on a line $x = 1/2$, the line along the middle of the critical strip."
Further tests by the think-tanks (la Vallée-Poussin, Hadamard, Hardy)
Showed that this is true for the first 100 billion zeros only.

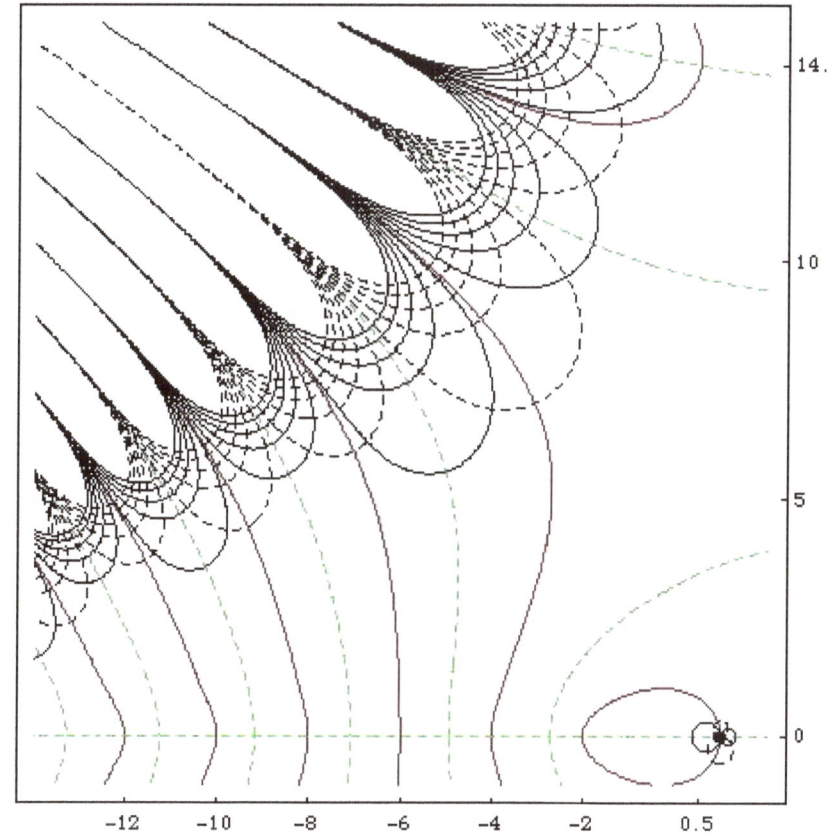

In this zeta function graph
Level curves for real $\zeta(s)$
are shown in solid lines;
The red curve indicates
$\zeta(s) = 0$,
The black curves represent
values other than zero.
Level curves for $Im\zeta(s)$ are
shown in dotted lines;
The green curve is when
$Im\zeta(s)=0$,
The function $\zeta(s)$ is real on
the real axis, where
$Im\zeta(s) = 0$.

Zeros of $\zeta(s)$ are points where both $\zeta(s)$ and $Im\zeta(s)$ are 0.
These are points where the red and green curves cross.
Trivial zeros are seen at the negative even integers,
And the first non-trivial zero at the coordinate point { 1/2, 14.135 }.

Put somewhat differently,
The harmonic series is a special case for the zeta function $\zeta(s)$.
And the real valued zeta function given for r and n is this:

$$\xi(s) = \frac{s}{2}(s-1)\pi^{-\frac{s}{2}}\Gamma(\frac{s}{2})\zeta(s)$$

$$\zeta(n) = \sum_{r=1}^{\infty}\frac{1}{r^n} = 1 + \frac{1}{2^n} + \frac{1}{3^n} + \frac{1}{4^n} + \cdots + \frac{1}{r^n}$$

So, if you put in for n = 1, you get the harmonic series that diverges.
For all values of n > 1 however, the series converges,
Meaning the sum tends towards some number as the value r increases,
Yet, it does not run off into infinity.

Last not the least, there is a fascinating connection
Between the Reimann zeta function $\zeta(x)$ and the prime numbers
2, 3, 5, 7, 11, 13, 17, 19, 23, ...
Forming the following expression that turns to be a zeta function:

$$(1 - 1/2^2) \times (1 - 1/3^2) \times (1 - 1/5^2) \times (1 - 1/7^2) \cdots$$

Which some name as a "composite ladder."

In summary:
Since its publication (1896), several proofs have been found
For the Reimann paper,
Including elementary proofs by Selberg and Erdós.
However, Riemann's hypothesis about the roots of zeta function
Still remains a mystery as of today (07.17.17).

In 1900, when David Hilbert set out his famous 23 problems
He declared his central inquiry, "if I were to sleep for 500 years
Then to awaken, my first question would be:
Has the Reimann hypothesis been broken?"